WJEC

GCSE

WORKBOOK

Physics

Jeremy Pollard

Higher tier

Suitable for Physics GCSE and Science (Double Award) GCSE

Answers online

HODDER
EDUCATION
LEARN MORE

Contents

WORKBOOK

1 This workbook will help prepare you for the WJEC GCSE Physics and WJEC GCSE Double Award (Physics) exams.

2 Your exams will include a range of short, structured questions and longer questions requiring continuous prose, including 6-mark questions testing both your knowledge of physics and the quality of your written communication. You need to be able to answer questions assessing knowledge and understanding, solve chemistry problems using mathematical skills and analyse and evaluate data. This workbook will help you develop the skills to answer all these question types. All questions are suitable for Higher Tier students. For both types of questions, those relating to the Science Double Award are shown in the paler tint and those for physics only in the darker tint.

3 Included are:
- stimulus materials, including key terms and concepts
- short and longer exam-style questions
- space for you to write your answers

4 Answering the questions will help you develop your skills and meet the assessment objectives AO1 (knowledge and understanding), AO2 (application) and AO3 (analysis and evaluation).

5 You still need to read your textbook and refer to your revision guides and lesson notes.

6 Timings are given for the exam-style questions to make your practice as realistic as possible.

7 Marks available are indicated for all questions so that you can gauge the level of detail required in your answers.

8 Answers are available at:
www.hoddereducation.co.uk/workbookanswers

9 When asked to 'select an equation ...', refer to the table, which is at the web address above.

Unit 1: Electricity, energy and waves

1.1 Electric circuits

In this topic you need to be able to describe and use the relationship between current (I) and potential difference (pd or voltage, V) and their relationship to the concept of resistance (R), using the equation:

$$I = \frac{V}{R}$$

This is often expressed in term of an I–V graph, called an electrical characteristic. You need to know the shape of I–V characteristics for a fixed resistor, a filament bulb and a diode. You also need to know the patterns of how voltage and current are related to each other in series and parallel circuits. In a series circuit, the current is the same throughout the circuit and the voltages add up to the supply voltage. In parallel circuits, the voltage is the same across each branch of the circuit and the sum of the currents in each branch is equal to the current in the supply. The total resistance (R_T) of resistors in series is equal to the sum of all the individual resistances. For two resistors (R_1 and R_2) in series:

$$R_T = R_1 + R_2$$

and for two resistances in parallel:

$$\frac{1}{R_T} = \frac{1}{R_1} + \frac{1}{R_2}$$

The electrical power (P) of a component is defined as the energy transferred (E) per unit time (t) and can be calculated using the equations:

$$E = Pt, \ P = VI \text{ and } P = I^2R$$

① **An electrical engineer is designing the circuits for a new mains-powered hairdryer. The hairdryer has a motor to power the fan, two ceramic heating resistors; a thermistor to monitor the temperature of the air coming out of the hairdryer and a variable resistor that can be used to control the temperature of the air.**

   a   **The two ceramic heating resistors are connected to each other in parallel, and they are in turn connected in series with the variable resistor and the power supply. Draw a suitable circuit for this part of the hairdryer.** `3 marks`

   b   **On the HIGH heat setting, the variable resistor has a resistance of 410 Ω. The voltage across the variable resistor is 130 V. Select and use a suitable equation to calculate the current.** `3 marks`

         ....

         ....

         ....

①① **These questions are for GCSE Physics students only**

c Also on the HIGH setting, the two ceramic heating resistors have a resistance of 180 Ω each. Select and use a suitable equation to calculate the total resistance of the two ceramic resistors in parallel.   3 marks

....

....

....

d Select and use a suitable equation to calculate the total resistance of the two ceramic resistors in series with the variable resistor on the HIGH setting.   2 marks

....

....

Exam-style questions   35

1 a Below are four *I–V* graphs (A, B, C and D) showing the electrical characteristics of some components.

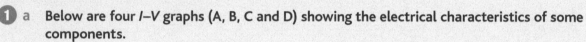
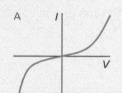
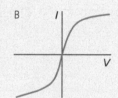
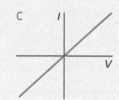
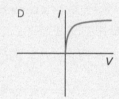

Select the correct letter for each of the following components: 3 marks

i   fixed resistor   *I–V* graph: ....

ii   diode   *I–V* graph: ....

iii   filament bulb   *I–V* graph: ....

b An electrical engineer wants to examine the behaviour of a ceramic heating resistor inside a hairdryer over a range of different currents and voltages. She wants to add into its circuit a filament indicator bulb to show when the hairdryer is working properly. She connects both the components, one at a time, into a circuit together with an ammeter, a voltmeter and a variable power supply unit to determine the voltages over a range of different currents.

i   Draw a circuit diagram to show how the electrical engineer could do this for the filament bulb.   3 marks

1 **1** These questions are for Science Double Award and GCSE Physics students

ii A table of the electrical engineer's results is shown below:

		Current (A)	0.0	0.2	0.4	0.6	0.8	1.0
Voltage (V)	Bulb		0.0	100	140	160	170	180
	Resistor		0.0	40	80	120	160	200

Use this data to plot an electrical characteristic graph. Include both components on the same graph.

4 marks

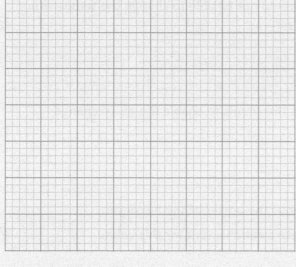

c Describe how the resistance of the filament bulb changes with the current flowing through it.

4 marks

....

....

d The circuit diagram below is set up by the electronic engineer to control the fan motor, M, and the two ceramic heating resistors inside a hairdryer.

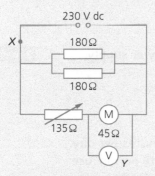

230 V dc

X

180 Ω

180 Ω

135 Ω

M

45 Ω

V

Y

Use this diagram to calculate:

i The voltage, Y, across the motor

3 marks

....

....

....

....

....

....

ii   The total current, *X*, drawn from the power supply

4 marks

....

....

....

iii Suggest a reason why this circuit would not be able to change the temperature of the air coming out of the hairdryer.

2 marks

....

....

iv   Suggest a suitable improvement to the circuit to enable the air to change temperature.

2 marks

....

....

e   The electrical engineer needs to work out the power of the hairdryer.

i   Calculate the power of the fan. State a suitable unit for this value.

3 marks

....

....

....

ii   If the current flowing through each of the ceramic resistors is 1.3 A, calculate the power of each of the ceramic resistors.

3 marks

....

....

iii Hence calculate the total electrical power drawn from the power supply.

2 marks

....

....

iv   The hairdryer is intended to be used for a maximum of 10 minutes per day. If it is used on the HIGH setting for the whole of the 10 minute period, use your answer to part **iii**. to calculate the electrical energy used by the hairdryer each day.

3 marks

....

....

....

1.2 Generating electricity

This topic covers the advantages and disadvantages of renewable and non-renewable technologies for the generation of electrical power in terms of efficiency, reliability, carbon footprint and power output. You need to be able to discuss the need for the National Grid as a nationwide electrical distribution system and the use of step-up and step-down transformers in the transmission of electricity from the power station to the home.

1 1 These questions are for Science Double Award and GCSE Physics students

The law of conservation of energy can be applied to different situations, such as investigating data comparing the efficiency of power stations and explaining why transmitting energy from power stations at high voltage and low current is an efficient way of transferring energy. You also need to be able to describe the overall redistribution of energy in a system using Sankey diagrams; and to apply the relationship between power, voltage and current, $P = VI$, to calculate the current flowing when electrical power is transmitted at different voltages. Finally, you need to be able to use and apply the equation for efficiency in terms of energy and power:

$$\%\text{efficiency} = \frac{\text{energy [or power] usefully transferred}}{\text{total energy [or power] supplied}} \times 100$$

1 Holy Island, off Anglesey, is the second largest island in Wales. The island is connected to Anglesey by two road causeways and a bridge, and the main town of Holyhead is a major ferry port connecting North Wales to Ireland. The vast majority of Holy Island's electrical power is brought onto the island by the National Grid via pylons. A small amount of renewable energy is generated locally, but 8 km offshore from the island is Holyhead Deep, the location for a proposed 80 MW commercial tidal power station. 12 km from Holyhead town is the former nuclear power station at Wylfa, currently being decommissioned, and the nearest fossil fuel power stations, such as Connah's Quay, are 80 km away on the River Dee. Holy Island is also part of the Anglesey Area of Outstanding Natural Beauty.

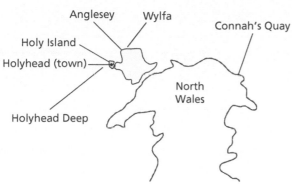

a Energy sources can be described as either *renewable* or *non-renewable*. Define the words in italics. **2 marks**

  i renewable means:

  ....

  ....

  ii non-renewable means:

  ....

  ....

b The Welsh Government has a policy of increasing the amount of electrical energy generated using renewable means so that Wales can bring about a lower carbon-dependent future. The graph below shows the amounts of electrical energy generated from different sources between 2004 and 2016

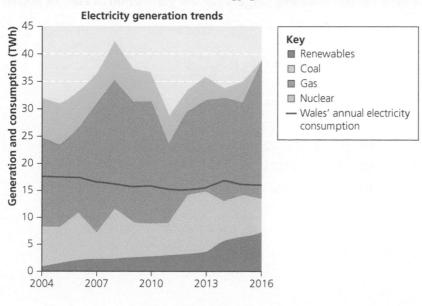

i Describe the trend in the generation of electrical energy from renewables from 2004 to 2016, and how this trend could affect the future carbon footprint of electrical energy generation in Wales.

3 marks

....

....

....

ii Identify and suggest possible explanations for the trends in electrical energy generation from 2004 to 2016 from the following energy sources:

4 marks

Nuclear trend:

....

....

Suggested explanation:

....

....

Coal trend:

....

....

Suggested explanation:

....

....

Exam-style questions

15

1 In 2010, the Welsh Government set the following targets for the generation of electrical energy by renewable means:

Type of renewable energy technology	Target capacity (GW)	Target date
Onshore wind	2.0	2017
Offshore wind	6.0	2016
Biomass	1.0	2020
Tidal	8.5	2022
Wave	4.0	2025
Local photovoltaic	1.0	2020

1 **1** These questions are for Science Double Award and GCSE Physics students

i   Plot a chart on the grid below showing the target capacity (GW) against the type of renewable energy technology: `3 marks`

ii   Explain why other tidal power stations will be needed in addition to the Holyhead Deep Tidal Power Station in order to meet the Welsh Government's target for tidal power generation by 2022.   `2 marks`

....

....

....

b   The Welsh Government is considering a proposal for a new nuclear power station at Wylfa on Anglesey. It has the potential to generate 2700 MW of electrical energy. The output voltage of the power station will be 25 000 V. Use this information and a suitable equation to determine the output electric current from the proposed power station at Wylfa. Give a suitable unit for this value.   `4 marks`

....

....

Output current = ... Unit = ...

c   Connah's Quay Power Station is a 1420 MW gas-fired power station in Flintshire in north Wales, next to the River Dee. It operates at 45% overall efficiency.

i   Use this data and a suitable equation to determine the total energy input as chemical stored energy in the gas supplied to the power station.   `3 marks`

....

....

....

Total chemical power supplied = ... MW

ii   Explain why the electrical energy generated from Connah's Quay is transmitted in the National Grid at *high* voltages and *low* currents.   `3 marks`

....

....

....

1 1 These questions are for GCSE Physics students only

1.3 Making use of energy

Thermal (heat) energy will transfer from somewhere hot (at a higher temperature) to somewhere colder (at a lower temperature) by the mechanisms of conduction, convection and radiation. A molecular model of matter can be used to explain the differences in these mechanisms.

During conduction, the vibration of particles, connected by bonds, transfers the vibrations from hot to cold. Metals are better conductors than non-metals due to the presence of mobile electrons in their structure.

Convection in liquids and gases can be explained in terms of molecular movement from hot to cold. Hotter liquid or gas molecules move faster than colder molecules and are further apart from each other (increasing their volume) and so hotter liquids or gases are less dense than colder ones. Hotter, less dense liquids or gases float above colder, more dense liquids or gases. The density of a material is calculated using the equation:

$$\text{density} = \frac{\text{mass}}{\text{volume}}$$

$$\rho = \frac{m}{v}$$

The particle model of matter can be used to explain the different properties and behaviours of solids, liquids and gases, – for example, the particles in a solid are closer together than in a liquid or a gas, so solids (of the same material) are denser than their liquids or their gases.

Thermal radiation involves the emission of infra-red electromagnetic waves from hot objects. Dull, black surfaces are good emitters and absorbers of thermal radiation, and shiny, silvered surfaces are good reflectors of thermal radiation.

1 In this question you will investigate the payback times for several different insulation systems that could be fitted to a house. The diagram below shows a detached house with four possible insulation systems that could be fitted.

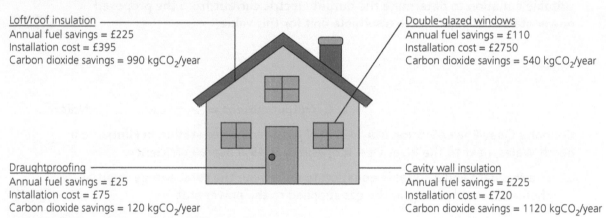

Loft/roof insulation
Annual fuel savings = £225
Installation cost = £395
Carbon dioxide savings = 990 kgCO$_2$/year

Double-glazed windows
Annual fuel savings = £110
Installation cost = £2750
Carbon dioxide savings = 540 kgCO$_2$/year

Draughtproofing
Annual fuel savings = £25
Installation cost = £75
Carbon dioxide savings = 120 kgCO$_2$/year

Cavity wall insulation
Annual fuel savings = £225
Installation cost = £720
Carbon dioxide savings = 1120 kgCO$_2$/year

The payback time for each system can be calculated using:

$$\text{payback time (years)} = \frac{\text{installation cost (£)}}{\text{annual savings (£/year)}}$$

a **Use the equation above and data from the diagram to complete the table.** `4 marks`

Insulation system	Payback time (years)
Loft/roof insulation	
Double-glazing	
Cavity wall insulation	
Draught excluders	

1 1 These questions are for Science Double Award and GCSE Physics students

b The householder is considering installing loft/roof insulation **or** cavity wall insulation. Complete the table below to compare the advantages and disadvantages of each system and suggest which one would be best for the householder. You should consider at least six different points; use capital letters and full stops in your answers. **8 marks**

Insulation system	Advantages	Disadvantages
Loft/roof insulation		
Cavity wall insulation		

Exam-style questions

15

1 A heating engineer is analysing the mechanism of heating a lounge. The lounge contains a large hot-water radiator fixed to one of the walls. The room is heated primarily by the convection of hot air around the room as shown in the diagram below.

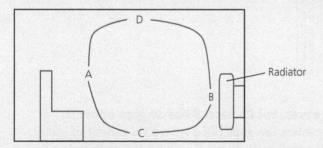

a Air circulates around the room via a convection current. Use the letters on the diagram to describe the direction of the convection current around the room, starting from the heat source. **1 mark**

1 1 These questions are for GCSE Physics students only

11

b The air circulates due to changes in the densities of hotter and colder air. When the heating system is OFF, the air inside the room has a mass of 78 kg. The dimensions of the room are 10 m × 3 m × 2 m.

   i   Calculate the volume of the room. `1 mark`

   ....

   ii Use a suitable equation to determine the density of the air when the heating system is OFF. State the unit. `3 marks`

   ....

   ....

                              Density = ... Unit = ...

   iii At which point, A, B, C or D, in the diagram above is the air MOST dense, when the heating system is ON. `1 mark`

   ....

   iv At which point, A, B, C or D, in the diagram above is the air HOTTEST, when the heating system is ON. `1 mark`

   ....

   v   Explain, using a molecular model, why the air circulates around the room via a convection current. `4 marks`

   ....

   ....

   ....

   ....

c The heating engineer suggests two designs for the radiator in the room, as shown in the diagrams below.

   Radiator A   Radiator B

   Both radiators have the same overall surface area, but Radiator B has air gaps between each of the heating tubes, whereas Radiator A has the gaps filled with a thin sheet of metal. Radiator A is painted white and Radiator B is a dull grey colour. Compare the radiators and explain why Radiator B is likely to be a more efficient way of heating the room. `6 marks`

   ....

   ....

   ....

   ....

1 1 These questions are for Science Double Award and GCSE Physics students

1.4 Domestic electricity

Domestic electricity needs to be both safe and cost-effective. In this topic you need to understand the functions of fuses and circuit breakers in preventing the flow of current when faults occur in household circuits. You need to be able to explain how fuses, miniature circuit breakers (mcb) and residual current circuit breakers (rccb) work and to calculate the appropriate current ratings of fuses.

Electricity is carried around a house via a ring main circuit consisting of a combination of live, neutral and earth wires. The live and neutral wires transmit the electricity, and the earth wire is a safety system providing a very low-resistance pathway for the electric current to flow along when there is a short circuit, reducing the possibility of electrocution.

Domestic electricity consumption is measured in kilowatt-hours (kWh) or units. 1 kWh is the amount of electrical energy used by an electric device with a power of 1 kW operating for 1 hour. Electricity companies charge consumers per unit of electrical energy consumed, where:

units used (kWh) = power (kW) × time (h) and cost = units used × cost per unit

---

1 Domestic appliances, such as toasters and kettles, consume large quantities of electrical energy. The current passing through them can be very high and potentially dangerous if a fault, such as a short circuit, occurs. These appliances connect to the house ring main using a fused plug. The fuse rating of the plug is chosen so that it is above the normal operating current of the appliance. Domestic fuses are generally only available in the following values: 3 A, 5 A or 13 A. The value of the fuse should be printed on the plug fitted to the appliance and printed on the side or base of the appliance itself, but older appliances may have standard general plugs and you may have to work out the value of the fuse that needs to be fitted. The table shows the power of selected mains electric devices. They all operate at 230 V. The operating current of the appliance can be determined using the equation $P = VI$.

Complete the table.

`12 marks`

Appliance	Power (kW or W)	Operating current (A)	Selected fuse (3 A, 5 A or 13 A)
Kettle	2.2 kW		
Microwave oven	900 W		
TV	450 W		
Hairdryer	1800 W		
Laptop computer	575 W		
Toaster	1 kW		

---

1 1 These questions are for GCSE Physics students only

Exam-style questions

1 The diagram below shows part of a ring main for a house.

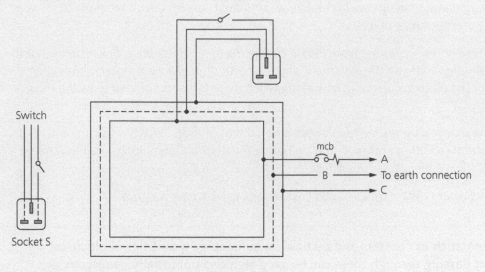

a Identify wires A, B and C, and their colours. 〔6 marks〕

A is the .... wire, coloured ....
B is the .... wire, coloured ....
C is the .... wire, coloured ....

b An electrician is fitting a new socket, S, into a bedroom. Complete the circuit diagram above showing how she should connect the socket to the ring main. 〔2 marks〕

c i The ring main has a miniature circuit breaker (mcb) fitted rated at 32 A. Explain the advantage of using an mcb compared to an older style ceramic fuse fitted with 32 A fuse wire. 〔2 marks〕

....

....

ii Mains voltage in the UK is 230 V. Identify a suitable equation and use it to calculate the maximum electrical power that could be drawn from this ring main before the mcb switches off. What is the correct unit for this value. 〔3 marks〕

....

....

Maximum power = ... Unit = ...

d The electrician notices that the ring main does not have a residual current circuit breaker (rccb). State the two wires that the rccb should be connected across and explain how an rccb works. 〔3 marks〕

The rccb is connected between the .... and the .... wires on the ring main.

....

....

....

1 1 These questions are for Science Double Award and GCSE Physics students

2 The Welsh Government provides a feed-in tariff for homeowners who generate their own electricity using photovoltaic (PV) solar panels and small wind turbines. This is an annual payment of £240 in addition to any money made exporting unused electricity via the National Grid. The table below compares two commonly available systems.

	4 kW Photovoltaic solar panels	6 kW pole-mounted wind turbine
Purchase and installation costs (£)	£8000	£21 000
Energy saving per year (£)	£160	£1250
Carbon dioxide saved (kgCO$_2$/year)	1590	4500
Annual feed-in tariff (£)	£240	£240
Payback time (years)		

a When calculating the payback time of a system, the annual feed-in tariff needs to be taken into account. Explain how the payback time is calculated and complete the table.

`3 marks`

....

....

b A householder is considering installing one of these systems. The house is situated on the edge of a small coastal town; it has a large garden and a south-facing roof. Compare the advantages and disadvantages of the two systems and suggest and explain which system the householder should choose. You should ensure that your answer takes into account the quality of your written communication (QWC).

`6 marks`

....

....

....

....

....

....

....

....

....

1 1 These questions are for GCSE Physics students only

1.5 Features of waves

In this topic you need to understand the basic properties of both transverse and longitudinal waves and the differences between them. You need to be able to describe these in terms of amplitude, wavelength (λ), frequency (f) and wave speed (v); and you need to be able to label these quantities in a diagram of a transverse wave.

Wave speed, frequency and wavelength are related to each other by the wave equation $v = f\lambda$, and the speed of a wavefront is given by the equation:

$$\text{wave speed} = \frac{\text{distance}}{\text{time}}$$

When waves reflect they obey the law of reflection, and when they move from one side of a refracting boundary to another they change speed, undergo refraction and change wavelength.

The electromagnetic spectrum is an important family of transverse waves consisting of different regions, which, in order of decreasing wavelength, are radio waves, microwaves, infra-red, visible light, ultraviolet, X-rays and gamma rays. Electromagnetic waves all travel at the same speed in a vacuum, the speed of light, but they have different wavelengths and frequencies. Radio waves have the longest wavelengths (and lowest frequencies), and gamma rays have the shortest wavelengths (and highest frequencies).

Electromagnetic waves and the (nuclear) energy emitted by radioactive materials are both referred to as 'radiation', but only nuclear radiation, ultraviolet, X-rays and gamma ray electromagnetic radiations are ionising, meaning that they are able to interact with atoms and damage cells by the energy that they carry.

All regions of the electromagnetic spectrum transfer energy, and certain regions (such as radio waves, microwaves, infra-red and visible light) are commonly used to transmit information. Radio and microwaves are used in communications using satellites in geosynchronous (or geostationary) orbits.

---

1 A student does an experiment to measure how the speed of water waves in a plastic tray varies with the depth of the water, from 5 mm to 35 mm in 5 mm increments, as shown in the diagram below.

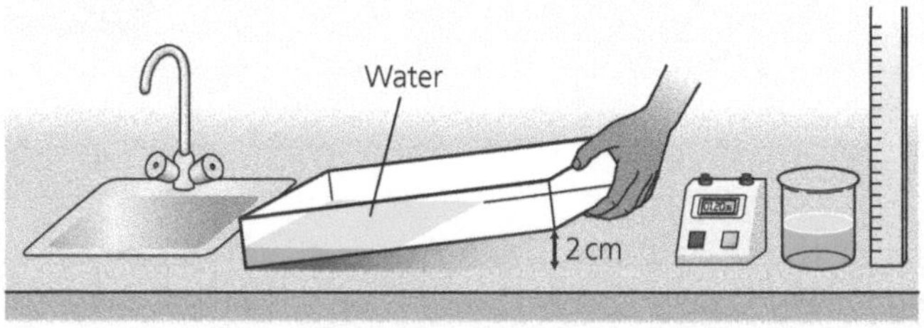

Water

2 cm

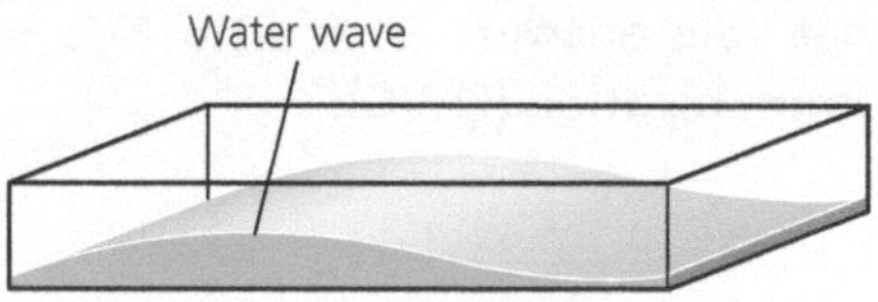

Water wave

a   Write a method for this experiment.   5 marks

....

....

....

....

....

---

1 1 These questions are for Science Double Award and GCSE Physics students

b   The student obtains the data in the table from this experiment. The water wave travels a total distance of 0.76 m.

Depth of water, d (mm)	Time, t (s)				Mean speed, v (m/s)
	1	2	3	Mean	
5	3.24	3.18	3.16		
10	2.25	2.23	2.28		
15	1.90	2.92	1.91		
20	1.75	1.74	1.71		
25	1.68	1.67	1.65		
30	1.62	1.63	1.61		
35	1.57	1.58	1.59		

   i   One of the readings in the table is an anomalous result. Identify this anomaly.   **1 mark**

   ....

   ii   Suggest a reason for this anomaly.   **1 mark**

   ....

   ....

   iii   Ignoring this anomaly, calculate the mean time for the wavefront to travel up and down the tray.   **7 marks**

   iv   Hence calculate the mean speed of the wave.   **7 marks**

c   On the graph paper below, plot a graph of mean speed (y-axis) against depth of water (x-axis), and draw a line of best fit.   **4 marks**

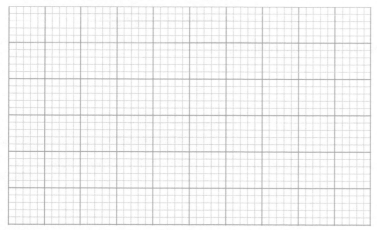

d   Describe the pattern in the results.   **2 marks**

   ....

   ....

e   Use your graph to estimate the mean speed of the waves with a depth of water equal to 2.5 mm.   **1 mark**

   ....

❶ ❶ These questions are for GCSE Physics students only

f The student assesses that her results are repeatable. Using examples from the table, state and explain whether you agree with her assessment or not. `2 marks`

....

....

....

Exam-style questions

20

1 Water waves are generated inside a ripple tank by a horizontal bar that moves up and down in the water 6 times in 8 seconds, generating water waves with a vertical distance of 0.8 cm from peak to trough. The waves then take 20 s to travel from one end of the ripple tank to the other, a total distance of 60 cm.

a What is the amplitude of the water waves? Give a suitable unit for your measurement. `3 marks`

....

....

Amplitude = ... Unit = ...

b Select a suitable equation and use it to calculate the speed of the water waves in the ripple tank. Express your answer in metres per second, m/s. `3 marks`

....

....

Wave speed = ... m/s

c Calculate the frequency of the water waves in Hertz, Hz. `3 marks`

....

....

Frequency = ... Hz

d Select and use a suitable equation to calculate the wavelength of the water waves, in metres, m. `3 marks`

....

....

....

Wavelength = ... m

e The horizontal bar is then adjusted so that it moves in and out of the water 12 times in 8 seconds. Circle the correct words to complete the following sentences describing the changes to the water waves: `4 marks`

i   The frequency of the water waves will (double / halve / stay the same).

ii   The speed of the water waves will (double / halve / stay the same).

iii The wavelength of the waves will (double / halve / stay the same).

iv   The amplitude of the waves will (double / halve / stay the same).

1 1 These questions are for Science Double Award and GCSE Physics students

f A flat plastic plate is put into the ripple tank water so that the water waves travel in shallower water as they pass over the plate as shown below.

Deep water

Shallow water

Plastic plate

State the name of the physical process that occurs when the water waves travel from deep water into shallow water and describe the changes that happen to the waves. **4 marks**

....

....

....

....

1.6 The total internal reflection of waves

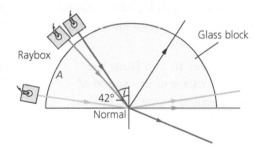

Raybox

Glass block

A

42°

Normal

Total internal reflection of light occurs when the light rays travel across a boundary between two transparent media, such as glass into air. Light travels slower in dense transparent materials such as glass, water and plastic, and faster in air or in a vacuum. The light will refract as it crosses the boundary and, because it speeds up, if it hits the boundary at an angle, the ray will appear to bend away from the normal line (an imaginary line at right angles to the boundary) as shown by the red line in the diagram below.

If the angle of incidence to the boundary is bigger than a certain 'critical' angle (which depends on the transparent materials), then the angle of refraction will be greater than 90° and the light ray will reflect off the boundary, obeying the law of reflection, and undergo total internal reflection, as shown by the green line in the diagram.

Optical fibres use this property of light (and infra-red) to transmit signals at high speeds over long distances. They can also be manufactured to be much thinner than human hairs, and many different wavelengths of the light or infra-red can travel along a single fibre at the same time, resulting in the ability to transmit huge amounts of information along a cable containing many thousands of fibres. Optical fibre communications can reduce the time delays in telephone or video conversations over very long distances compared to signals sent as microwaves via geostationary satellites. Optical fibres are also very flexible and have been put to good use inside medical instruments, such as endoscopes, that can allow internal medical examinations without using ionising radiation, such as X-rays, and keyhole surgery.

1 In this question you will compare the medical imaging techniques of endoscopy and CT scans. A medical endoscope usually has two sets of optical fibres inside it. One set of fibres takes light from a source down through the endoscope and another set picks up the light that is reflected off the inside of the body and transmits it back up the endoscope, so it can be displayed on a screen for the doctor. The endoscope tube is flexible and can be inserted into the body through a suitable orifice (the mouth or the anus), or a small incision ('keyhole'). The diagram shows the end of a typical endoscope.

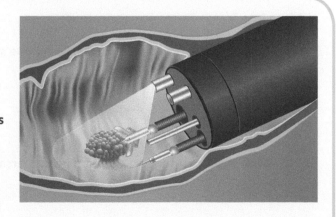

1 1 These questions are for GCSE Physics students only

Medical CT scans involve using a beam of X-rays that pass through parts of the body and are detected on the other side. The beam is slowly rotated and moved around the part of the body, and a computer imaging system collates all the X-ray signals and converts them into a high resolution 3D image that can be manipulated on a screen.

Match the following statements to the correct medical imaging technique. Some statements may apply to both techniques.

10 marks

Uses ionising radiation	
Produces a high-resolution image	
Biopsy samples can be obtained	Endoscopy
Real time images can be taken	
A non-invasive technique	
Requires a trained operator	CT scanning
Operator needs to be protected	
Uses total internal reflection	

Exam-style questions

1 A businesswoman in Miami, USA (at A in the diagram below) wants to have a telephone call with her business partner in Perth, Western Australia (at B). The telephone signal can go via a satellite system in geosynchronous orbit or via an undersea/overland optical fibre link.

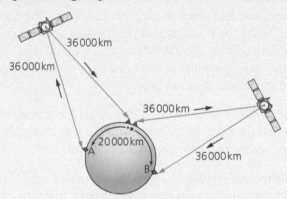

36 000 km
36 000 km
36 000 km
20 000 km
A
B
36 000 km

The speed of microwaves to and from the satellites is 3.0×10^5 km/s and the speed of the infrared rays travelling along the optical fibres is 2.0×10^5 km/s.

a Calculate the total distance travelled by the microwaves from A to B via the two satellites. **2 marks**

....

....

....

b Use a suitable equation to determine the time it takes for a telephone signal to travel from Miami to Perth via the satellite system. **3 marks**

....

....

Time = ... s

c Calculate the time taken for the telephone signal to travel via the optical fibre link. **2 marks**

....

....

Time = ... s

d Compare the effect on a telephone conversation of sending signals via each system. **2 marks**

....

....

1 1 These questions are for Science Double Award and GCSE Physics students

1.7 Seismic waves

Seismic waves are generated when slabs of rock move relative to each other. This can happen between the Earth's tectonic plates, or as the result of volcanic activity, or they can be generated artificially using explosions or very large hammers.

There are three main types of seismic waves. P (or primary) waves are the fastest waves and travel at 5–8 km/s. They are longitudinal in nature and can travel through both the solid and liquid regions of the Earth's interior. S (or secondary) waves travel at slower speeds than P waves, typically 2–8 km/s, and they are transverse in nature and cannot travel through the liquid regions of the Earth's interior. The third type of seismic waves are surface waves that only propagate across the Earth's surface and are the slowest seismic waves travelling at typical speeds of 1–6 km/s.

Seismic waves can be detected using a device called a seismometer. By comparing the arrival times of P and S waves, the distance of the epicentre of a disturbance (usually an earthquake) from a seismometer station can be calculated. Combining the signals from at least three widely spaced stations can pinpoint the position of the earthquake epicentre.

Analysis of many thousands of seismograms has led to the current model of the large-scale structure of the interior of the Earth. The diagram below shows the passage of S (Diagram 1) and P (Diagram 2) seismic waves through the Earth following an earthquake.

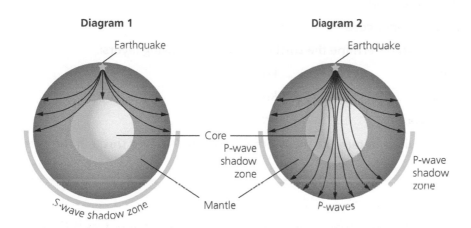

As there is an S-wave shadow zone on the opposite side of the Earth from the earthquake and S-waves cannot propagate through liquids, so the Earth must have a solid mantle and (at least a partially) liquid iron core. Careful analysis of the passage of P waves through the Earth's core has led to evidence that the inner core is made from solid metallic iron.

---

1 The table shows the speed of P-waves at different depths through the Earth.

Depth (km)	Speed of P-waves (km/s)
0	5.0
100	7.5
2800	13.0
2900	7.5
5000	12.5
5500	12.5

---

1 1 These questions are for GCSE Physics students only

a Draw a graph of this data. 〔4 marks〕

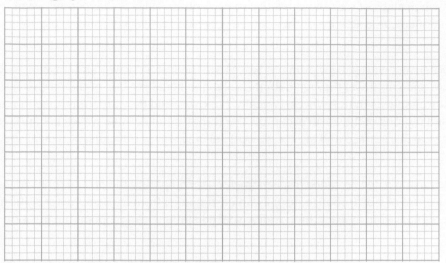

b Label on your graph the following boundaries: 〔3 marks〕

   i crust – mantle

   ii mantle – outer core

   iii outer core – inner core

c Use your graph to determine the thickness of the following layers: 〔3 marks〕

   i the crust .... km

   ii the mantle .... km

   iii the outer core .... km

d **i** Use the graph to determine the average speed of P-waves through the mantle. 〔2 marks〕

....

....

   ii Hence estimate the mean time it would take for a P-wave to travel through the mantle from the bottom of the crust to the top of the inner core. 〔2 marks〕

....

....

Exam-style questions

1 On 23 May 1975, a magnitude 3.8 earthquake hit South Wales. The earthquake was recorded by three seismograph stations in Wales – one on Anglesey, one in Pembroke and one in Monmouth. The combined seismograms, showing the arrival of the seismic waves at each of the three stations, is shown in the diagram.

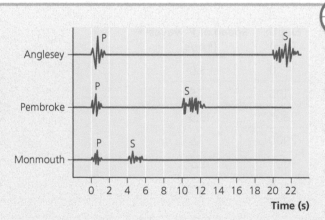

1 1 These questions are for Science Double Award and GCSE Physics students

a Use the seismogram to measure the S–P seismic wave time-lag (in s) between the arrival of both types of seismic wave at each of the three stations. Record your values in the table.

`3 marks`

Station	S–P time-lag (s)	Distance to epicentre of the earthquake (km)
Anglesey		
Pembroke		
Monmouth		

b The distance of each seismometer station from the epicentre of the earthquake is given by the equation:

$$\text{distance (km)} = \left(\frac{\text{S–P time-lag}}{5}\right) \times 60$$

Use this equation to complete the table, calculating the distance of each seismogram station from the epicentre.

`3 marks`

c The map below is an outline of Wales, showing the positions of the three seismometer stations, use the scale and data from the table above, to determine the approximate epicentre of the 1975 earthquake.

`4 marks`

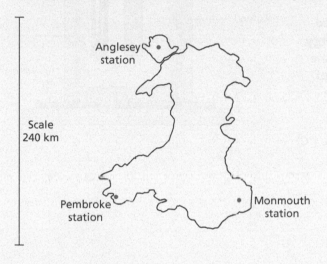

1.8 Kinetic theory

In this topic you need to understand the concept of pressure in terms of a force acting over a given area, given by the equation $p = \dfrac{F}{A}$.

The pressure, p, volume, V, and the temperature, T, of a fixed mass of gas are interconnected to each other by the equation: $pV/T = \text{constant}$, where p will be measured in N/m^2 (or Pa), V will be measured in m^2 and T is the absolute temperature of the gas in kelvin, K. The absolute (or Kelvin) temperature scale is based on the concept of absolute zero, which is the temperature at which molecular motion of the gas particles ceases, $-273°C$ or 0 K. A temperature in kelvin can be calculated using the equation: $T_{kelvin} = T_{celsius} + 273$.

A more useful version of the pV/T equation allows you to calculate changes in pressure, volume or temperature from one set of conditions (1) to another set of conditions (2), given by:

$$\frac{p_1 V_1}{T_1} = \frac{p_2 V_2}{T_2}$$

1 **1** These questions are for GCSE Physics students only

23

The kinetic theory model can be applied to the gas – increasing the temperature of the gas particles increases their speed and hence the pressure that they exert on the container walls. If the volume of the gas is allowed to change (like in a balloon), then it will increase as well.

You also need to know the equations that relate heat transfer to changes in temperature and state. A fixed mass of material, m, will change its temperature by $\Delta\theta\degree C$, when a quantity of heat energy, Q, is transferred to it, as given by the equation $Q = mc\Delta\theta$, where c is the specific heat capacity of the material. Transferring an amount of heat energy, Q, to a mass, m, of a material at its melting/boiling point resulting in a change of state given by $Q = mL$, where L is the latent heat of the material.

1 The specific heat capacity, c, of a solid material can be determined by heating a known mass of the material using an electric heater, and measuring the resultant temperature change. The most common apparatus for doing this is shown in the diagram.

Lagging the solid block reduces the heat energy that is transferred to the surroundings, so that most of the electrical energy supplied by the heater causes the block to heat up. A student performed this experiment on a copper block with a mass of 0.5 kg. She used a joulemeter to measure the electrical energy supplied to the heater, and she measured the temperature of the block for every 1000 J of electrical energy supplied to the block. Her results are shown in the table.

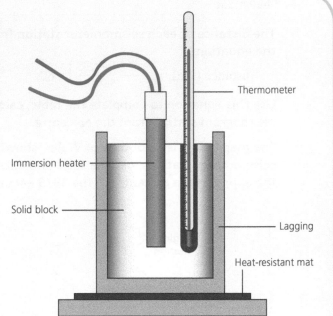

Thermometer

Immersion heater

Solid block

Lagging

Heat-resistant mat

Electrical energy supplied (J)	Temperature (°C)
0	20
1000	25
2000	30
3000	36
4000	41
5000	46
6000	51
7000	56

a Use her data to determine the specific heat capacity of the copper block. **6 marks**

Specific heat capacity of the copper block, c = ... J/kg°C

....

....

b The student looks on the internet and finds that the given value of the specific heat capacity of copper is 385 J/kg°C. Suggest reasons why the value that you have worked out from the student's data is different to the value given on the internet. **2 marks**

....

....

1 1 These questions are for Science Double Award and GCSE Physics students

Exam-style questions

1 A keen cyclist is pumping up the tyres on his mountain bike. He notices that if he seals the end of the pump and compresses the air inside, the changing pressure causes a change in volume inside the pump. He records the changes in the table:

Pressure, p (atmospheres)	Volume, V (cm³)
1.0	48
1.5	32
2.0	24
2.5	19
3.0	16

a i On the graph paper below, plot a graph of volume, V (y-axis) against pressure, p (x-axis).   **3 marks**

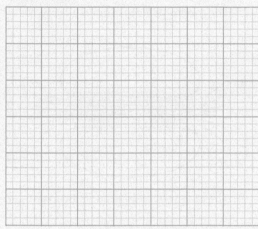

ii Describe the variation of the volume with the pressure.   **2 marks**

....

....

....

....

iii Use the graph to estimate the volume of the air inside the pump when the pressure is 3.5 atm pressure.   **1 mark**

....

b The cyclist prefers to reduce the air pressure in the tyres of his bike when he is cycling over rough ground. Suggest and explain why this is a good idea.   **2 marks**

....

....

c At the start of a road race, when the temperature of the air in the tyres is 20°C, the cyclist pumps his tyres up to a pressure of 2.5 atm. The volume of air inside the tyre is 1800 cm³. At the end of the race, the temperature of the air in the tyres has risen to 40°C, but the volume has not changed.

i Convert the temperature at the start and finish of the race to degrees kelvin.   **2 marks**

20°C is .... K

40°C is .... K

ii Calculate the pressure of the air in the tyre at the end of the race.   **3 marks**

....

....

....

Pressure at the end of the race = ... atm

1 **1** These questions are for GCSE Physics students only

1.9 Electromagnetism

Magnetic fields are places where magnetic materials experience a force. They can be created by permanent magnets (made out of iron, cobalt or nickel), or by an electric current flowing through a wire, or a coil, called an electromagnet. The magnetic field patterns produced by a permanent magnet, a current-carrying wire, a solenoid (coil) and a current loop are shown in the diagrams below.

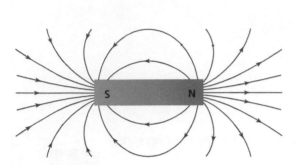

The magnetic field around a bar magnet.

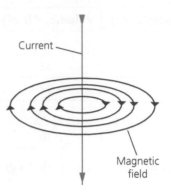

The magnetic field around a current-carrying wire.

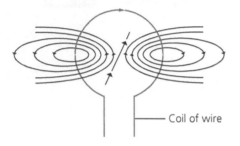

The magnetic field around a current-carrying solenoid.

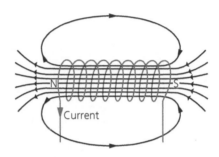

The magnetic field around a current-carrying coil.

When an electric current flows through a wire inside a magnetic field, the wire and the magnetic field exert a force on each other that can make either of them move – called the motor effect. Fleming's left-hand rule can be used to predict the direction of the force on the conductor, the current or the magnetic field, as shown here.

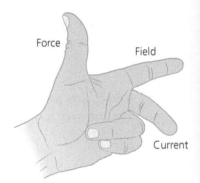

The size of the force, F, experienced by a current-carrying conductor of length l, inside a magnetic field of strength B, at right angles to the current, I, is given by $F = BIl$.

Electromagnetic induction works in the opposite sense to the motor effect, whereby a moving wire or a changing magnetic field can induce a current to flow in the wire. This effect is used to make electric generators and the output from the generator can be increased by moving (or spinning) the conducting wire faster' increasing the size of the magnetic field or increasing the number of turns on the coil. The relative directions of the induced current, movement of the wire and the magnetic field are given by Fleming's right-hand rule, as shown here.

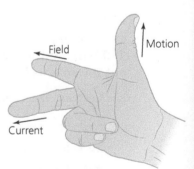

An alternating voltage, V_1, across a primary coil (with N_1 turns), surrounding an iron core, can induce an alternating voltage, V_2, in a secondary coil (with N_2 turns). If the transformer is 100% efficient then:

$$\frac{V_1}{V_2} = \frac{N_1}{N_2}$$

1 A teacher shows the action of a transformer to her class by setting up the experiment shown in the diagram below.

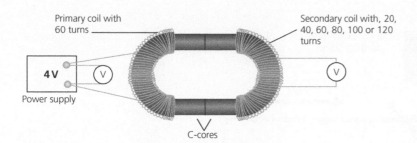

The graph of her results is shown below.

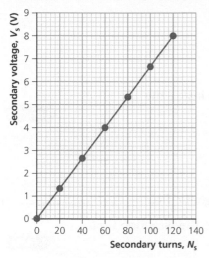

a In this experiment, identify: **5 marks**

i the independent variable ....

ii the dependent variable ....

iii a control variable ....

iv the range of the independent variable ....

v the range of the dependent variable ....

b The teacher explains to the class that her results show that the transformer equation, $\dfrac{V_1}{V_2} = \dfrac{N_1}{N_2}$ is correct. Use data from the graph and the diagram to confirm the teacher's claim.   **4 marks**

....

....

....

....

1 1 These questions are for GCSE Physics students only

Exam-style questions

1 A copper wire carries a current of 0.6 A through a magnetic field of strength 1.6 T, as shown in the diagram below.

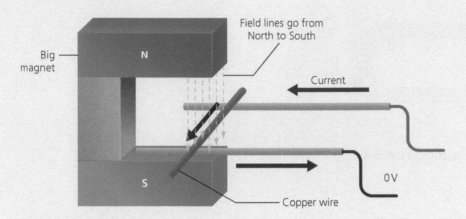

The force acting on the wire is 0.0385 N.

a Use a suitable equation to determine the length of the wire inside the magnetic field. Give your answer in centimetres, cm.

4 marks

....

....

Length of wire inside the magnetic field = ... cm

b The force acting on the copper wire causes it to move. Circle the direction that the copper wire moves:

1 mark

towards the back of the magnet or towards the power supply

2 An simple electric motor is shown in the diagram below. A battery is connected into the green base, which passes a current through the coil of wire. The coil then spins inside the magnetic field.

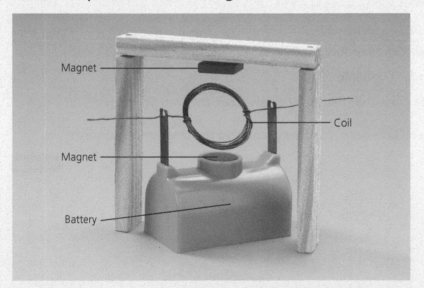

What changes can be made to this design that will increase the speed of rotation of the coil?

3 marks

....

....

....

1 1 These questions are for Science Double Award and GCSE Physics students

Unit 2: Forces, space and radioactivity

2.1 Distance, speed and acceleration

Speed is a measure of how fast an object is moving. The speed of an object can be defined as the distance that an object moves in a given time, and it can be calculated using the equation:

$$\text{speed} = \frac{\text{distance}}{\text{time}}$$

The velocity of an object is its speed in a given direction.

When objects change speed they are said to accelerate if they get faster, and decelerate if they go slower. The acceleration (or deceleration) of an object is given by the equation:

$$\text{acceleration} = \frac{\text{change in velocity}}{\text{time}}$$

The movement of objects can be analysed using graphs of motion. The gradient (or slope) of a distance–time graph is the speed of the object, and a horizontal line on the graph indicates a stationary object, as shown in the graph below.

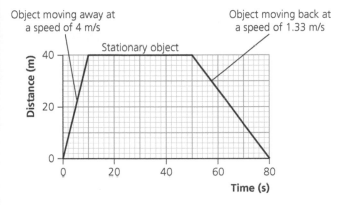

The velocity–time graph of the motion of an object shows more information than a distance–time graph. The gradient (or slope) of the graph gives the acceleration of the object, and the area under the graph shows the distance travelled by the object.

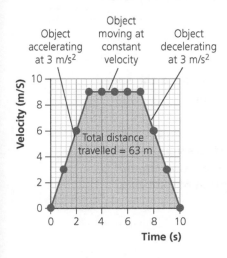

The overall stopping distance of a vehicle depends on the thinking distance (the distance travelled by the vehicle during the reaction time of the driver) and the braking distance (the distance travelled by the vehicle during the time that the brakes are being applied).

overall stopping distance = thinking distance + braking distance

The thinking distance depends on: the speed of the vehicle; reaction time of the driver (this is affected by drowsiness, alcohol and drugs); driver distractions such as passengers or in-car technologies such as mobile phones.

🅖🅖 These questions are for GCSE Physics students only

Braking distance depends on the speed and mass of the car, the condition of the brakes and the road and the weather.

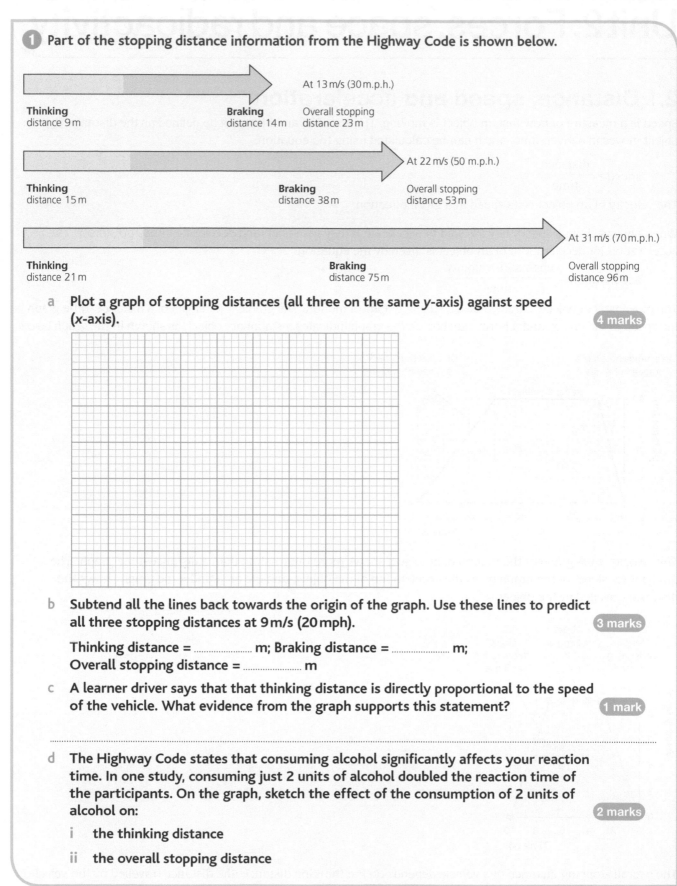

1 **Part of the stopping distance information from the Highway Code is shown below.**

At 13 m/s (30 m.p.h.)

Thinking distance 9 m   **Braking** distance 14 m   Overall stopping distance 23 m

At 22 m/s (50 m.p.h.)

Thinking distance 15 m   **Braking** distance 38 m   Overall stopping distance 53 m

At 31 m/s (70 m.p.h.)

Thinking distance 21 m   **Braking** distance 75 m   Overall stopping distance 96 m

a **Plot a graph of stopping distances (all three on the same *y*-axis) against speed (*x*-axis).** `4 marks`

b **Subtend all the lines back towards the origin of the graph. Use these lines to predict all three stopping distances at 9 m/s (20 mph).** `3 marks`

Thinking distance = ... m; Braking distance = ... m; Overall stopping distance = ... m

c **A learner driver says that that thinking distance is directly proportional to the speed of the vehicle. What evidence from the graph supports this statement?** `1 mark`

....

d **The Highway Code states that consuming alcohol significantly affects your reaction time. In one study, consuming just 2 units of alcohol doubled the reaction time of the participants. On the graph, sketch the effect of the consumption of 2 units of alcohol on:** `2 marks`

    i **the thinking distance**

    ii **the overall stopping distance**

1 1 **These questions are for Science Double Award and GCSE Physics students**

Exam-style questions

15

1 An athlete runs a 400 m race in 50 s. Calculate her average speed during this race.

`1 mark`

....

2 A car is travelling at 9 m/s (20 mph). Calculate the time taken to travel 40.5 m.

`1 mark`

....

3 A velocity–time graph for a drone is shown below.

Choose suitable equations to calculate:

a The initial acceleration of the drone and state the unit.

`3 marks`

....

....

....

Acceleration = ... Unit = ...

b The distance travelled during the time that the drone was travelling at constant velocity.

`3 marks`

....

....

....

Distance travelled at constant velocity = ... m

c The total distance travelled during the flight.

`5 marks`

....

....

....

....

Total distance travelled = ... m

d The drone takes off once more and accelerates to 20 m/s as in part **a**. The drone then accelerates at 2.5 m/s² for a further 3 seconds until it is flying at its maximum velocity. Calculate the maximum velocity of the drone.

`3 marks`

....

....

Maximum velocity of the drone = ... m/s

1 **1** These questions are for GCSE Physics students only

2.2 Newton's laws

In this topic you will need to be able to define the concepts of inertia, mass and weight. Inertia is a measure of how easy or difficult it is to get something that is stationary to move, or to change its motion once it is moving. Inertia is defined as the resistance of any object to a change in its state of motion or rest. Mass is a measure of how much material there is in an object. Mass is measured in kilograms, kg. The mass of an object is the same, no matter where it is. Weight is the force of a gravitational field acting on the mass of an object. Weight depends on the mass, m, of the object and the gravitational field strength, g. Weight is measured in newtons, N. On Earth, the gravitational field strength, g, is approximately 10 N/kg. The weight of an object is given by the equation:

weight (N) = mass (kg) × gravitational field strength (N/kg)

Newton's first law of motion describes the inertia of an object:

'An object at rest stays at rest, or an object in motion stays in motion with the same speed in the same direction, unless acted upon by an unbalanced force.'

When an unbalanced (or resultant) force acts on an object it produces a change in the object's motion or it distorts (changes the shape) of the object. The resultant force will cause the object to accelerate (or decelerate). The acceleration of the objects is directly proportional to the resultant force and inversely proportional to the object's mass, as given by Newton's second law of motion, which states that:

resultant force = mass × acceleration or $F = ma$

Newton's third law of motion takes into account the fact that forces always act in pairs and it states:

'In an interaction between two objects, A and B, the force exerted by body A on body B is equal and oppositely directed to the force exerted by body B on body A.'

Or, in other words:

'For every action force, there is an equal and opposite reaction force.'

When using Newton's third law, the following points need to be understood:

1   The two forces in an interaction pair act on different objects.

2   The two forces are equal in size, but act in opposite directions.

3   The two forces are always the same type – for example contact forces or gravitational forces (action-at-a-distance forces).

Finally you need to know about the forces and their effects to explain the behaviour of objects moving through the air, where friction always acts, causing objects to reach a terminal speed when the weight of a falling object downwards is equal and opposite to the air resistance acting upwards on the object, as shown by the skydiver.

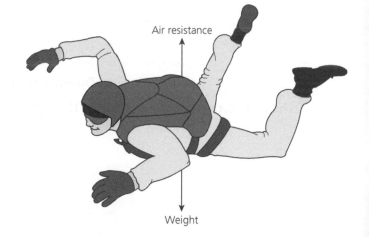
Air resistance

Weight

1 1 **These questions are for Science Double Award and GCSE Physics students**

1 In 1972 an experiment was proposed (but never carried out) by the crew of Apollo 17, the last manned mission to the Moon. The experiment was a simple one involving measuring the weights of carefully controlled masses on the surface of the Earth and the Moon, using specially calibrated newtonmeters. Modern satellite measurements have reproduced the data that would have been taken and are shown in the graph below.

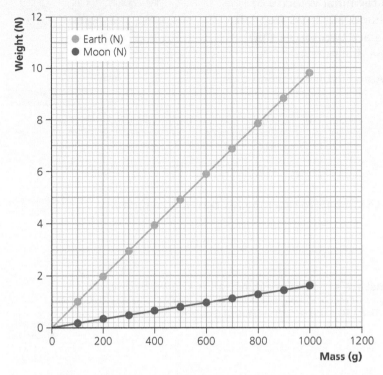

a   Use the graph above to calculate the gravitational field strength (take care with the units of mass):

   i   on the Earth

   ....

   ....

   ....

   ii   on the Moon

   ....

   ....

   ....

b   Explain why the weight of a 1 kg mass on the Moon is **less** than its weight the Earth. **2 marks**

   ....

   ....

   ....

c   If the experiment were to be repeated on Mars, the gravitational field strength would be 3.8 N/kg. On the graph above, sketch the graph that you would expect for Mars.

Exam style questions

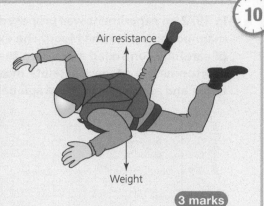

Air resistance

Weight

1 In 2012, the world record for free-fall parachuting was broken by Alan Eustace, a computer scientist from Google. He jumped out of a helium-filled balloon from a height of 41 419 m, and reached a terminal velocity of 367 m/s, faster than the speed of sound. His total free fall time was 52 s. During free-fall there are two forces that act on a parachutist, as shown in the diagram.

   a Using the diagram explain why Alan Eustace reached a terminal velocity of 36 m/s.

3 marks

....

....

....

   b Explain why he fell more than 37.5 km before reaching his terminal velocity.

2 marks

....

....

   c A velocity–time graph of Alan Eustace's record-breaking fall is shown below.

Explain how the graph shows that Alan Eustace reached a second terminal velocity of about 50 m/s.

4 marks

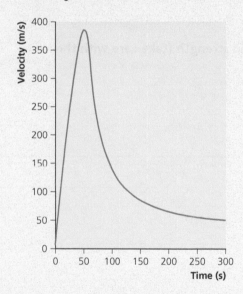

....

....

....

....

....

....

....

2.3 Work and energy

When a force acts on a moving object, energy is transferred, although the total amount of energy remains constant. The force is said to do work on the object:

    work done = force × distance in the direction of the force or $W = F \times d$

Objects possess energy if they are moving, and this is called kinetic energy. For an object of mass, m, moving at a speed, v, is:

$$\text{kinetic energy} = \frac{\text{mass} \times (\text{velocity})^2}{2} \text{ or } KE = \frac{1}{2}mv^2$$

Objects can also possess energy due to their position (height) inside a gravitational field - this is called gravitational potential energy. An object of mass, m, inside a gravitational field of strength, g, moved through a height difference, h, possesses gravitational potential energy, PE.

    change in potential energy = mass × gravitational field strength × change in height or $PE = mgh$

1 1 These questions are for Science Double Award and GCSE Physics students

Objects such as springs can be deformed and energy can be stored in them called elastic energy. When a force, F, is applied to a spring, it can cause the spring to extend by a distance, x, and the force is directly proportional to the extension, given by the equation:

force = spring constant × extension or $F = kx$

where k is the spring constant. Very stiff springs have high spring constants and require a lot of force to stretch them by a small extension. The work done stretching a spring is given by the area under a force (y-axis) versus extension (x-axis) graph. This work done, W, in stretching a spring is given by the equation:

work done = $\frac{1}{2}$ × force × extension or $W = \frac{1}{2}Fx$

The energy efficiency of vehicles is also important when considering the work done when they move. The more efficient a vehicle is, the less energy is wasted to the environment. The energy efficiency of a vehicle can be improved by streamlining the shape of the vehicle; making it easier for the air to pass around it as it is moving. Large amounts of energy are also wasted by rolling resistance between the vehicle's tyres and the road surface. Energy can also be recovered from electric vehicles when they are idling or moving downhill as the unused kinetic energy can drive a dynamo, recharging the battery.

The safety of drivers and passengers in vehicles is improved by fitting safety features such as seat belts, air bags and crumple zones. All three of these work by increasing the contact time of the collision when a vehicle is involved in an accident. Increasing the collision time reduces the forces acting on the occupants. Some of the kinetic energy of the collision is absorbed by safety features – stretching the seat belts, deforming the air bag or crumpling the engine or the boot of the vehicle.

1. A student performed a simple experiment stretching the spring of a baby-bouncer. He added masses to the baby harness and measured the extension of the spring. His results are shown in the table.

Weight (N)	Extension (m)			
	1st experiment	2nd experiment	3rd experiment	Mean
0.0	0.00	0.00	0.01	
5.0	0.01	0.01	0.02	
10.0	0.02	0.03	0.04	
15.0	0.03	0.04	0.05	
20.0	0.05	0.05	0.07	
25.0	0.06	0.06	0.08	
30.0	0.07	0.07	0.09	
35.0	0.09	0.08	0.11	
40.0	0.10	0.09	0.12	

a Suggest suitable pieces of equipment for measuring the weight of the masses **and** the extension of the spring.　　2 marks

....

....

b State the resolution of the two measuring instruments used to obtain the data in the table above.　　2 marks

....

....

1 1 These questions are for GCSE Physics students only

c State the dependent and independent variables and their ranges. **4 marks**

Independent variable = ... Range = ...

Dependent variable = ... Range = ...

d Complete the table by calculating the mean extension value for each weight – write your values to a suitable number of decimal places. **10 marks**

e Plot a graph of weight (*y*-axis) versus mean extension (*x*-axis). Draw a line of best fit for the data. **4 marks**

f Use your graph and a suitable equation to determine the spring constant of the spring. **3 marks**

....

....

....

....

....

....

....

....

g The student notices that when he performed the experiment for the third time, the extension of the spring had not returned to zero.

i Suggest a reason for this. **1 mark**

....

....

ii Is this a *measurement* error, a *systematic* error or a *random* error in the 3rd set of measurements? **1 mark**

....

iii Suggest a possible way that he could take this into account in his results. **1 mark**

....

....

1 1 These questions are for Science Double Award and GCSE Physics students

Exam-style questions

15

1 The world's fastest roller coaster is the Formula Rosso in Abu Dhabi, UAE. It is only 52 m high, but the coaster train is accelerated to its top speed in 5 seconds by a hydraulic launch system, which produces a release velocity similar to that of steam catapults on an aircraft carrier. The coaster cars accelerate to 28 m/s (62 mph) in 2 seconds – faster than most supercars. A simplified sketch of the first part of the ride is shown below:

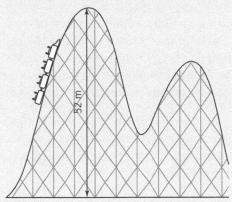

52 m

a Use a suitable equation to calculate the initial acceleration of the coaster car. `2 marks`

....

....

Acceleration = ... m/s²

b i Each of the coaster cars has a mass of 700 kg. Calculate the total mass of the roller coaster shown in the diagram above. `1 mark`

....

ii Select a suitable equation and calculate the gravitational potential energy of the roller coaster at the top of the ride. Assume g = 10 N/kg. `3 marks`

....

....

....

c The steam catapult 'fires' the roller coaster off the top of the ride, and it reaches a maximum speed of 67 m/s.

i Use a suitable equation to calculate the total kinetic energy of the roller coaster at its maximum speed. `3 marks`

....

....

ii Use your values from parts **ci** and **bii** to determine the kinetic energy transferred by the hydraulic launch system to the roller coaster. `2 marks`

....

iii State **one** assumption that you made when calculating in part **cii** `1 mark`

....

1 **1** These questions are for GCSE Physics students only

2.4 Further motion concepts

Rectilinear motion is motion in a straight line. The momentum of an object is a measure of its inertia, as it depends on its mass, m, and its velocity, v:

momentum = mass × velocity or $p = mv$

Newton's second law of motion can be written as:

change of momentum = force × time

Momentum is a quantity that is always conserved in the Universe. During any interaction between objects, the total momentum of all the objects before the interaction must equal the total momentum of all the objects after the collisions. This is called the law of conservation of momentum. Kinetic energy is not always conserved during collisions and explosions. If no energy is wasted in transfers to the environment, then the kinetic energy during a collision is conserved, but this does not happen normally because energy is wasted as heat, sound, light or deforming the object.

Momentum is also involved with Newton's third law of motion. If an object with a large mass collides with an object with a low mass, then each object exerts an equal and opposite force on the other. As the objects have different masses, they will move away from the collision with different accelerations.

Isaac Newton came up with four equations describing the rectilinear motion of objects. The following symbols are used in these equations:

- x – distance travelled (in m)
- u – initial velocity of the object (in m/s)
- v – final velocity of the object (in m/s)
- a – acceleration of the object (in m/s^2)
- t – time of the motion (in s).

The equations are:

- $v = u + at$
- $x = \dfrac{u + v}{2} t$
- $x = ut + \dfrac{1}{2} at^2$
- $v^2 = u^2 + 2ax$

When objects are able to rotate or move about a pivot, they do so because of moments. A moment is a turning force defined by the equation:

moment = force × distance (normal to the direction of the force) or $M = Fd$

The principle of moments involves systems with a pivot in balance, and it states that:

sum of the clockwise moments = sum of the anticlockwise moments

1 A student is investigating the principle of moments using the apparatus shown below.

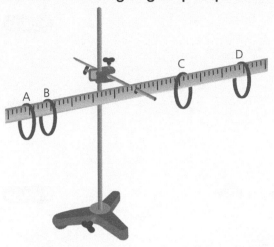

The student hangs slotted mass stacks from the loops at A, B, C or D and records the (weight) force and the distance from the pivot of each moment. Complete the table. All the forces in the table are recorded in newtons, N, and the distances are measured in cm, so the moments can be recorded in N cm. **2 marks**

| Anticlockwise | | | | | | Clockwise | | | | | | Motion – balanced/ rotates clockwise/ rotates anticlockwise |
| Moment A | | | Moment B | | | Moment C | | | Moment D | | | |
F	d	M	F	d	M	F	d	M	F	d	M	
2	25		–	–	–	5			–	–	–	balanced
3	10		2	15		1			1	50		balanced
	10	40		10	30	–	–	–	6		60	rotates anticlockwise
2		60	3	40		6	20		9		90	
4		32	2	9		5	4		2		20	

Exam-style questions

30

1 This question is about the sport of curling in which large 'stones' are pushed to slide along an ice rink towards a target at the far end. When curling stones hit each other they rebound, but a Welsh company has developed a new type of magnetic stone that can attach to another magnetic stone introducing a new variation on the game.

Two curling stones, A and B, are moving on an ice rink, as shown here.

Direction of positive velocities

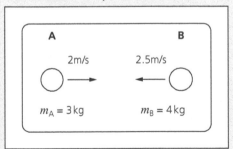

a State the unit of momentum. **1 mark**

....

1 **1** These questions are for GCSE Physics students only

b   Use a suitable equation to calculate:

   i   the initial momentum of stone A   `3 marks`

....

....

   ii   the initial momentum of stone B   `2 marks`

....

....

   iii   the total momentum of the stones before they collide   `2 marks`

....

c   The two standard stones rebound off each other. Stone A moves at a velocity
    of −1.8 m/s. Calculate the velocity of stone B.   `5 marks`

....

....

....

....

d   Two magnetic stones, C and D, are shown moving in the diagram below.

The two stones join together and move away from the collision point.
Calculate the velocity of the combined stones after the collision.   `10 marks`

....

....

....

....

....

....

1 **1** These questions are for Science Double Award and GCSE Physics students

2 A man is driving a pizza delivery moped. The moped is travelling at an initial velocity of 5 m/s before accelerating at 1.5 m/s^2 for 4 seconds.

    a Use suitable equations to calculate:

        i the final velocity of the moped   `3 marks`

....

....

        ii the distance travelled by the moped while it is accelerating   `3 marks`

....

....

    b The moped is moving at the velocity in part **a**, when it suddenly brakes to 2 m/s, travelling a distance of 10 m in order to go around a corner. Calculate the deceleration of the moped.   `3 marks`

....

....

....

2.5 Stars and planets

Our Solar System consists of eight planets (in roughly circular orbits away from the Sun – Mercury, Venus, Earth, Mars, Jupiter, Saturn, Uranus and Neptune); dwarf planets such as Pluto; over 180 moons; an asteroid belt; many long and short period comets; and our star, the Sun. The four inner planets are rocky or terrestrial planets, and the four outer planets are gas giants. Our galaxy, The Milky Way, is a huge collection of stars, all orbiting in a spiral pattern around a common galactic centre, thought to be a huge black hole.

Short distances, such as the distance from the Earth to the Moon, are measured in kilometres, but larger distances in the Solar System are measured in astronomical units (AU). Distances outside the Solar System are so large that astronomers use the distance that light travels in one year, called a light-year (l-y).

All stars form out of interstellar clouds of dust called nebulae. Gravity causes the dust to collapse in on itself, forming a protostar. When the temperature of the core of the protostar reaches 15 million °C, nuclear fusion starts and protostars become main sequence stars. On the main sequence the stability of a star depends on the balance between the gravitational force trying to pull it inwards and a combination of gas and radiation pressure trying to push it outwards.

Towards the end of their lifetime on the main sequence, Sun-like mass or lower stars run out of hydrogen and they start to fuse helium and other heavier elements, swelling out to form a red giant as they do so. The helium fusion era is relatively short and this eventually runs out as well, and the red giant collapses under its own weight. The outer atmosphere of the red giant is ejected back into space as a nebula and the remaining stellar core forms a white dwarf. This then cools and darkens with time. This is shown in the diagram below.

1 **1** These questions are for GCSE Physics students only

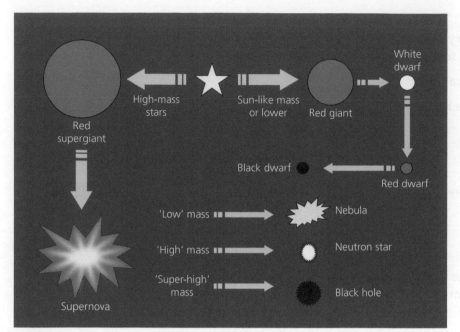

High mass stars have a much more violent death. As helium fusion starts, they swell up to form red supergiant stars. The resulting collapse of a red supergiant is so violent that it produces an enormous explosion called a supernova. Most of the red supergiant is ejected into space as a large nebula, but the remaining core depends on the initial mass of the star. Very high mass stars form neutron stars. The highest mass stars collapse with such a force that all the remaining material is effectively squashed into a tiny space forming a black hole.

The Hertzsprung–Russell (H–R) diagram was developed as a means of displaying the properties of stars, as shown in the diagram (right).

The H–R diagram has surface temperature (plotted backwards) on the *x*-axis and brightness on the *y*-axis. Stars move around on the diagram throughout their lifetime, following an evolutionary path.

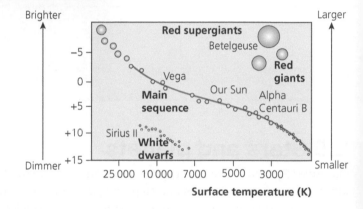

1 Use the diagram above, showing what happens at the end of a star's life-cycle, to describe and explain the stellar death sequence of:

a   the Sun   5 marks

b   Regulus (a very high-mass main sequence star in the constellation of Leo).   5 marks

Use the space below to plan your answer using bullet points, and then write your final answer on the lines below.

1 1 These questions are for Science Double Award and GCSE Physics students

....

....

....

....

....

Exam-style questions

10

1 The Hertzsprung–Russell (H–R) diagram shown above is a way of classifying stars.

a The diagram below is a blank H–R diagram.

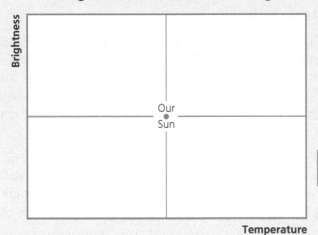

i On **both** the axes of the diagram, label them with the words **HIGH** and **LOW**. **2 marks**

ii Label the four quadrants of the diagram correctly using the following labels. Each quadrant should have one label: **4 marks**

Hot, bright stars	Cool, dim stars	Cool, bright stars	Hot, dim stars

b On the diagram below, draw a line showing the evolutionary path of the Sun.

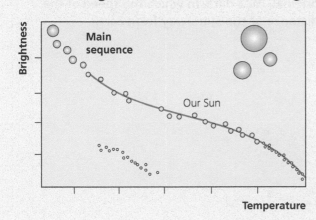

2.6 The Universe

This topic is about the Big Bang model of the formation and expansion of the Universe and the evidence that supports it. Scientists can analyse the light emitted by different elements at high temperatures by splitting the light up into its spectrum of different colours (or wavelengths). Each element has its own signature sequence of spectral lines. Large, extended, extremely hot objects, such as stars, emit a continuous spectrum containing all the colours and wavelengths, but as the light works its way out through the atmosphere of the star, elements in the atmosphere absorb some of the wavelengths that correspond to their signature sequence, forming an absorption spectrum.

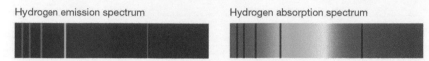

By analysing the pattern of absorption lines emitted by a star and collected by a telescope and spectroscope on Earth, the early astronomers could identify the composition of stars (by comparing the absorption line sequences to those produced by elements here on Earth).

When stars are moving away from or towards the Earth, the pattern of absorption lines produced by the star changes. The motion of the star either increases all the absorption wavelengths if it is moving away from Earth (called red shift); or it decreases all the absorption wavelengths if it is moving towards Earth (called blue shift). Only the closest stars to Earth appear to be blue shifting, and the vast majority of stars are red shifting leading to the conclusion that most stars (and galaxies) are moving away from Earth.

Spectroscopic observations and measurements of distant galaxies, made by Edwin Hubble in 1923, revealed that the wavelengths of the absorption lines emitted by the galaxies were red shifted and that the effect increases with distance away from Earth; the further away that you look, the faster the Universe is moving away from you. The observations can be explained by an expansion of the Universe. As the light moves away from a distant galaxy, so the space through which it is moving expands, increasing the wavelength of the light. This is called cosmological red shift.

Running cosmological red shift backwards implies that at some time in the distant past, the Universe was much smaller than it is now, and that at some initial time it was concentrated at a single explosive point. This model of the Universe and how it came into being became known as the Big Bang theory, and the remnants of the huge explosion that created the Universe at the time of the Big Bang can be observed today in the cosmic microwave background radiation that appears throughout the Universe.

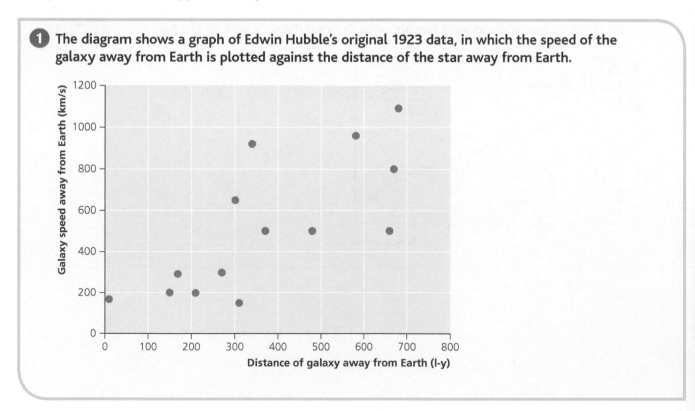

1 The diagram shows a graph of Edwin Hubble's original 1923 data, in which the speed of the galaxy away from Earth is plotted against the distance of the star away from Earth.

1 1 These questions are for Science Double Award and GCSE Physics students

a   Assuming, as Hubble did, that there is a linear relationship between these two variables, and that the graph starts at (0, 0) – the position of our own galaxy (The Milky Way) – plot a line of best fit through this data on the diagram above.   **1 mark**

b   Explain why many of Hubble's peers were very sceptical of his measurements and conclusions.   **2 marks**

....

....

....

c   A modern version of Hubble's graph is shown in the diagram below. This uses a modern data set, using galaxies that are very large distances away from Earth.

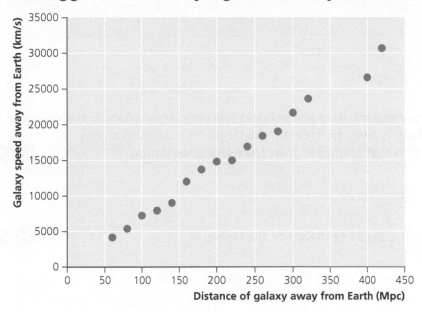

The unit of distance in this graph, Mpc or megaparsecs, is considerably larger than the l-y. 1 Mpc is equal to 3.26 million l-y.

i   Plot a line of best fit on the modern data in the graph above.   **1 mark**

ii   Explain where the Hubble data set would be on the graph above.   **2 marks**

....

....

iii   The modern data set is much more convincing and is now widely accepted by scientists. What is the relationship, known as Hubble's law, that can be deduced from the graph above?   **2 marks**

....

....

🔵🔵 These questions are for GCSE Physics students only

Exam-style questions

1 This question is about evidence for the Big Bang theory.

a Explain the difference between an 'emission spectrum' and an 'absorption spectrum'. **2 marks**

....

....

....

b The absorption spectra of light from distant galaxies show cosmological red shift. Explain what is meant by the term 'cosmological red shift'. **5 marks**

....

....

....

....

c The light emitted by very distant stars and galaxies has very large values of cosmological red shift. The further a galaxy is away from Earth, the larger its cosmological red shift. State the conclusion that scientists draw from this observation. **2 marks**

....

....

d State what is meant by the Big Bang theory of the Universe. **3 marks**

....

....

....

....

e Explain why the Cosmic Microwave Background Radiation (CMBR) is observational evidence for the Big Bang theory. **3 marks**

....

....

....

....

....

....

....

1 **1** These questions are for Science Double Award and GCSE Physics students

2.7 Types of radiation

The nuclei of some atoms are unstable due to an imbalance between the numbers of protons and neutrons. The nuclei can become more stable by the process of radioactive decay.

The common notation used to describe nuclei is in the form of $^A_Z X$, where X is the chemical symbol for the element, Z is the proton number and A is the nucleon number (the number of protons + the number of neutrons). Isotopes are nuclei of the same element that have the same number of protons (Z) but different numbers of neutrons, and hence different nucleon numbers (A).

During alpha particle decay, two protons and two neutrons are ejected from the nucleus, joined together forming a helium nucleus, $^4_2 He^{2+}$. These are sometimes written as $^4_2 \alpha$. Americium-241 is an alpha particle emitter. The decay equation for this isotope is:

$$^{241}_{95} Am \rightarrow {}^{237}_{93} Np + {}^4_2 \alpha$$

Alpha particles are the least penetrating of the radioactive particles, being absorbed by a sheet of paper.

Beta particles are high-energy electrons emitted from a nucleus when a neutron decays into a proton and an electron. A common beta emitter is strontium-90. Its decay equation is:

$$^{90}_{38} Sr \rightarrow {}^{90}_{39} Y + {}^{\ 0}_{-1} e$$

Beta particles are more penetrating than alpha particles and they are absorbed by a few millimetres of aluminium.

Gamma radiation involves electromagnetic radiation in the form of high-energy gamma rays (γ). Gamma ray emission involves the nucleons inside a nucleus rearranging themselves into a lower energy configuration. Gamma radiation is the most penetrating form of radiation and most gamma rays are stopped by a few centimetres of lead or many metres of concrete.

The waste materials from nuclear reactors is highly radioactive, ionising and dangerous to living material, and will remain so for many thousands of years. This means that it needs to be stored securely inside lead and steel containers in concrete-lined deep underground storage facilities.

Radioactive decay is a random process. It is not possible to predict exactly which unstable nucleus will decay, but every radioactive isotope has a (constant) probability of decay, which means that it is possible to predict how many will decay in a given time period. Experimental measurements of radioactive decay also need the background radiation to be taken into account. There are two groups of background radiation sources: naturally occurring radiation (such as the radon gas emitted by granite rocks, cosmic rays from space, radiation emitted by food and drink and gamma rays emitted by the ground and buildings); and artificial background radiation (such as that used in medicine, or from the nuclear industry, bomb fallout or accidents). A mean value of the background radiation needs to be taken, and this value is then subtracted from each measurement.

1 The table below shows the proportion (as the angle of a pie chart) of the daily count rate contribution of different sources of background radiation on the far west coast of Pembroke (where the underlying geology is mostly granite rocks that emit radon gas) and Aberystwyth (where the underlying geology is mostly sedimentary rocks).

Source of background radiation	Angle of pie chart (°)	
	Pembroke	Aberystwyth
Radon	285	180
Buildings and soil	22.5	60
Cosmic rays	22.5	45
Food and plants	15	45
Hospitals	7.5	15
Nuclear industry	7.5	15

1 1 These questions are for GCSE Physics students only

a On the pie chart outlines below, plot the data for each location. Each outline segment of the pie chart corresponds to 15°.

Pembroke

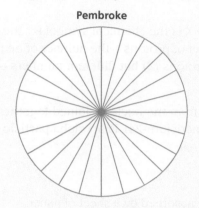

Aberystwyth

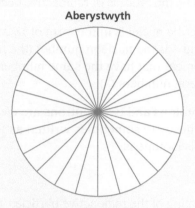

b Compare and contrast the background radiation in these locations.

4 marks

....

....

....

....

Exam-style questions

18

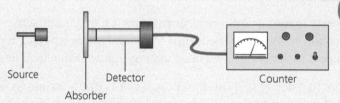

Source   Absorber   Detector   Counter

1 The apparatus shown on the right can be used to investigate the types of radiation emitted by a radiation source.

A student performed the experiment by placing different absorbers between the source and the detector and recording the counts in one minute (cpm). She measured and recorded the count rate without the source in place first, and repeated this measurement three times and calculated the mean value to be 90 cpm. Her results are shown in the table.

Absorber	Count rate (cpm)	Adjusted count rate (cpm)
None	620	
Sheet of paper	554	
3 mm thickness aluminium sheet	550	
2 cm thickness lead sheet	92	

a Explain why the student measured and recorded the count rate without the source first.

2 marks

....

....

b Complete the table.

1 mark

c The source appears to give out more than one type of radiation.

   i Name one type of radiation that is **not** given out by this source.

1 mark

....

1 1 These questions are for Science Double Award and GCSE Physics students

ii   What percentage of the emissions of the source is absorbed by the paper?   **2 marks**

....

....

iii   State and explain the effect of the lead absorber.   **2 marks**

....

....

d   The source is replaced by a pure gamma emitter, and the student places different thicknesses of lead between the source and the detector. Her adjusted measurements are shown in the table below.

Thickness of lead absorber (cm)	Adjusted count rate (cpm)
0	16 000
0.5	8 000
1.0	2 000
1.5	1 000
2.0	500

i   Plot a graph of these results on the graph paper below, and draw a line of best fit.   **4 marks**

ii   Use your graph to determine the 'half-thickness' of lead for these gamma rays. This is the thickness of lead required to halve the count rate.   **2 marks**

....

....

iii   State the relationship between the count rate and the thickness of the lead absorbers.   **2 marks**

....

....

....

iv   Use the graph to determine the proportion of the gamma rays that are absorbed by a 75 mm thick lead absorber.   **3 marks**

....

....

....

1 **1**   These questions are for GCSE Physics students only

2.8 Half-life

Radioactive decay is a random process with a constant probability of decay, and it can be modelled using any system that has a constant probability of change. Two common systems are dice (where each number on the dice has a $\frac{1}{6}$ probability of being thrown) or coins (where each side, heads/tails, has $\frac{1}{2}$ probability of being thrown). As an example, throwing 120 dice will result in 20 dice falling with a '1' upwards (representing decay) and throwing the remaining 100 dice will result in 16 or 17 '1' falling upwards, and so on until they have all 'decayed'. Radioactive decay is measured in becquerels, Bq, where 1 Bq is defined as 1 radioactive decay per second.

The constant probability of decay is usually expressed in terms of the half-life of a radioactive substance. This is the time taken for the initial activity of the radioactive decay to halve.

Radioactive substances have an extraordinary range of half-lives. Some radioactive substance created inside nuclear reactors can have incredibly short half-lives – hydrogen-7, for example, has a half-life of 23×10^{-24} s. But others can have extremely long half-lives – tellurium-128 has a half-life of 69×10^{30} s.

The half-life of a substance can be determined experimentally by plotting a decay curve. The graph below shows the radioactive decay curve of iridium-192.

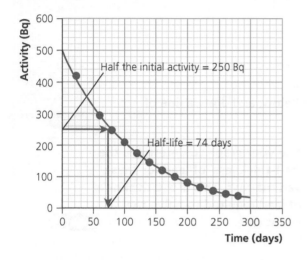

To measure the half-life, the initial activity is identified first, 500 Bq in this case. The half-life is then the time taken for the activity to halve, 250 Bq for this decay. Draw a line across from 250 Bq until it touches the decay curve and then subtending the line downwards gives the half-life on the time axis, 74 days for iridium-192.

---

1 A student is comparing dice and coins as models of radioactive decay. In his model, a six-sided dice is considered to have 'decayed' if it falls with a '1' upwards and a coin is considered to have decayed if it falls with a 'head' upwards. He starts with 480 dice and 480 coins. His data is shown in the table.

Number of throws	Number of DICE undecayed	Number of COINS undecayed
0	480	480
1		
2	333	120
3	278	60
4	231	30
5	193	15
6	161	8
7	134	4
8	112	2
9	93	1

1 **1** These questions are for Science Double Award and GCSE Physics students

a State the probability of decay of any one:

  i   dice = ...   `1 mark`

  ii coin = ...   `1 mark`

b Predict the numbers of each of the dice and the coins that will decay as a result of the first throw.

Dice : predicted number decayed after the first throw ....   `1 mark`

Coins : predicted number decayed after the first throw ....   `1 mark`

c Complete the table of the results, adding your predicted values for the first throw and then draw a graph of the student's results. Add lines of best fit to your graph.   `4 marks`

d Use your graph to determine the 'half-life' of the dice and the coins.   `2 marks`

e Predict the number of dice remaining after 10 throws.   `3 marks`

Exam-style questions

10

1 *Piltdown Man* is a famous scientific hoax that was eventually proved to be a fraud by carbon dating. In 1912, amateur archaeologist Charles Dawson claimed to have discovered human-like skull fragments and a jaw with two teeth, in the Piltdown gravel pit in Sussex. Dawson's discovery was analysed by Arthur Smith Woodward from the British Museum and they announced that the skull and the jaw belonged to a primitive human ancestor, who lived between 500 000 and 1 million years ago. The discovery was celebrated as the missing link between apes and humans, confirming Darwin's theory of evolution.

1 1 These questions are for GCSE Physics students only

During the 1930s the discovery of early hominids in Africa started to cast doubts on Piltdown Man's authenticity and, in 1939, the newly invented technique of fluorine dating concluded that the remains were no older than 50 000 to 500 000 years old. In 1959, carbon dating was used to date the remains. The graph below shows the radioactive decay curve for carbon-14.

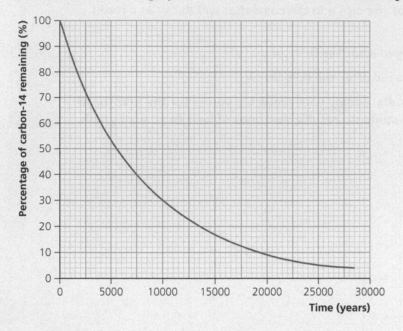

a   Use the graph to determine the half-life of carbon-14.   `2 marks`

b   Carbon-14 decays to nitrogen-14. Write the radioactive decay equation for
this change.   `3 marks`

c   The carbon-14 dating technique measured the percentage of carbon-14 remaining in the
samples to be between 90 and 95%. Use this data and the graph to determine the
range of ages for the Piltdown Man remains, and determine a mean value for the
age of the samples.   `3 marks`

d   If the Piltdown Man remains were genuine, use the graph to estimate the
percentage of carbon-14 remaining in them.   `1 mark`

e   Modern DNA analysis of Piltdown Man in 2009 have shown that the remains
are a mixture of a late medieval human and a modern orang-utan. Suggest why
it was difficult to scientifically authenticate the remains in 1912.   `1 mark`

1 1 These questions are for Science Double Award and GCSE Physics students

2.9 Nuclear decay and nuclear energy

Nuclear fission is the break-up of large nuclei into smaller 'daughter' nuclei with an accompanied release of energy. This energy can be used inside a nuclear reactor to generate heat and subsequently electricity. Many large 'fissile' nuclei can undergo nuclear fission, but uranium-235, $^{235}_{92}$U, is the most common fuel used in nuclear reactors – the process also emits neutrons that can induce the fission of other nuclei in a chain reaction.

Inside a nuclear fission reactor, the neutrons, $^{1}_{0}$n, emitted from spontaneous fission events are slowed down by moderator materials, such as water in order to increase the chance of further nuclear fissions. The slow-moving neutrons are absorbed by the uranium-235 nuclei and cause them to break up into two smaller daughter nuclei and a number of extra neutrons.

The rate of the subsequent chain reaction is adjusted by control rods made of boron, which absorb neutrons. These are moved up and down inside the reactor core, absorbing more or fewer neutrons. If the control rods are dropped to the bottom of the reactor, the whole system shuts down. During the fission process, some of the mass of the nucleus is transformed into energy, which can be used to heat water. This can then produce steam, which turns a turbine-generator system, generating electricity.

Nuclear fusion is the amalgamation of two light nuclei to form a heavier nucleus. This is accompanied by the release of energy as some of the mass involved with the fusion process is transformed into energy. Nuclear fusion of isotopes of hydrogen, $^{1}_{1}$H (including deuterium, $^{2}_{1}$H, and tritium, $^{3}_{1}$H), is the process that occurs within the core of stars. Nuclear fusion only happens at very high pressures and temperatures, and as such it is extremely difficult to produce controlled nuclear fusion on Earth. Experimental fusion reactors on Earth mostly use an immense doughnut-shaped magnetic field to confine the fusion material (such as deuterium and tritium), which is then heated to over 100 million °C.

The large numbers of highly energetic neutrons produced during nuclear fusion are a big safety problem for this type of reactor, as they require thick shielding to prevent their release into the surroundings.

---

1 This activity is about comparing the amount of energy released by the combustion of coal, the fission of uranium-235 and the fusion of hydrogen.

a   1 carbon atom has a mass of 1.99×10^{-26} kg, and the combustion of 1 carbon atom produces 6.47×10^{-19} J of heat energy.

i   Calculate the number of carbon atoms in 1 kg of carbon (coal). `1 mark`

ii   Use your value from **part i** to calculate the heat energy produced by the combustion of 1 kg of coal. `2 marks`

b   1 atom of uranium-235 has a mass of 3.90×10^{-25} kg, and emits 3.20×10^{-11} J of energy during nuclear fission.

i   Calculate energy released by the nuclear fission of 1 kg of U-235. `2 marks`

ii   The 260 MW Uskmouth B coal power station near Newport burns approximately 700 tonnes of coal per hour of operation (where 1 tonne = 1000 kg). Calculate the mass of uranium-235 required to generate the same amount of electrical energy as Uskmouth B power station. `2 marks`

1 **1** These questions are for GCSE Physics students only

c   The nuclear fusion of 1 kg of hydrogen produces 0.60×10^{15} J of energy.
Calculate the mass of hydrogen needed to produce the same amount of electrical
energy as Uskmouth B power station.

1 mark

....

....

d   Complete the table below comparing coal, uranium-235 and hydrogen as fuels
for Uskmouth B power station.

9 marks

Fuel source	Coal	Uranium-235	Hydrogen
Mass of fuel needed to produce 260 MW of electricity (kg)			
Safety concerns			
Environmental concerns			

Exam-style questions

(25)

1 These diagrams show a nuclear power station and a conventional coal-fired power station.

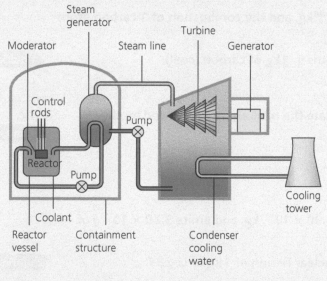

Nuclear power station

1 1 These questions are for Science Double Award and GCSE Physics students

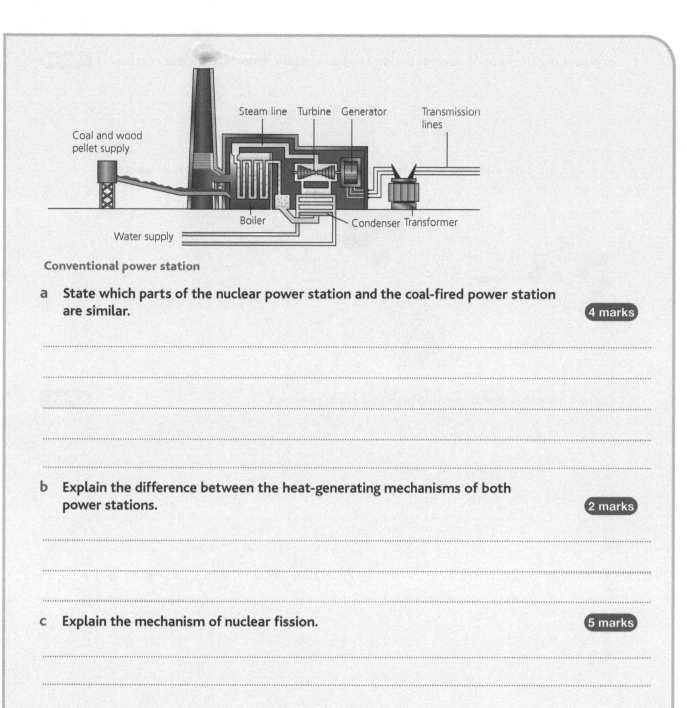

Conventional power station

a State which parts of the nuclear power station and the coal-fired power station are similar.

4 marks

....

....

....

....

b Explain the difference between the heat-generating mechanisms of both power stations.

2 marks

....

....

....

c Explain the mechanism of nuclear fission.

5 marks

....

....

....

....

....

d One nuclear fission reaction is shown below. Complete the equation.

3 marks

$$^{1}_{0}n + ^{235}_{92}U \rightarrow ^{\boxed{i}}_{54}Xe + ^{90}_{\boxed{ii}}Sr + 2\,^{1}_{\boxed{iii}}n + energy$$

i .... ii .... iii ....

e In the diagram of the nuclear power station, the moderator is pure water. Explain the function of a moderator.

2 marks

....

....

 These questions are for GCSE Physics students only

f   Explain how the rate of nuclear fission can be controlled inside a nuclear reactor.   **3 marks**

....

....

....

g   The diagram below shows a common nuclear fission reaction in a nuclear reactor.

$^{235}_{92}U$

$^{89}_{36}Kr$

Gamma ray

$^{1}_{0}n$

$^{1}_{0}n$

$^{1}_{0}n$ Impact by slow neutron

$^{1}_{0}n$

$^{144}_{56}Ba$

Explain how this reaction could lead to a chain reaction.   **4 marks**

....

....

....

....

Hachette UK's policy is to use papers that are natural, renewable and recyclable products and made from wood grown in well-managed forests and other controlled sources. The logging and manufacturing processes are expected to conform to the environmental regulations of the country of origin.

Orders: please contact Hachette UK Distribution, Hely Hutchinson Centre, Milton Road, Didcot, Oxfordshire, OX11 7HH.

Telephone: (44) 01235 827827.

Email education@hachette.co.uk

Lines are open from 9 a.m. to 5 p.m., Monday to Friday.

You can also order through our website: www.hoddereducation.co.uk

ISBN 978-1-5104-1904-9

© Jeremy Pollard 2018

First published in 2018 by

Hodder Education, an Hachette Company, Carmelite House, 50 Victoria Embankment, London, EC4Y 0DZ

www.hoddereducation.co.uk

Impression number   10 9 8 7 6 5

Year   2023

Cover photo: © Getty Images/iStockphoto/Thinkstock

Typeset by Aptara, India

Printed in Great Britain by Ashford Colour Press Ltd

A catalogue record for this title is available from the British Library.

HODDER EDUCATION

t: 01235 827827

e: education@hachette.co.uk

w: hoddereducation.co.uk

ISBN 978-1-5104-1904-9

MIX
Paper | Supporting responsible forestry
FSC™ C104740

9 781510 419049